Minecraft Halloween Tales:

A Collection of Five Spooky Stories

An unofficial spine-chilling Minecraft Book

By

DR. BLOCK

Be sure to check out my other books, including:

ISBN-13:978-1539663881
ISBN-10:1539663884

If you like this book, **please leave a review** so other Minecrafters can learn about it. Thank you!

Table of Contents

Talking Herobrine

“Come here, son!” I heard my dad yell.

I got up from my desk where I had been doing my homework. The life of a 12-year-old villager sure has a lot of homework in it. I walked out of my bedroom and went into the kitchen where my dad was standing with his hands behind his back.

“What is it, Dad?”

“Close your eyes.”

What am I, two years old? I didn't want to play peek-a-boo. Still, I indulged my dad and closed my eyes. "What is it?"

A few moments of silence passed and then my dad said, "Okay, open your eyes."

I opened my eyes and my dad was holding a small doll in his hands. Well, maybe I shouldn't call it a doll. That sounds kind of lame. I'll call it an action figure.

My eyes got wide when I realized what it was. "No way! You bought me a Talking Herobrine!"

That moment, my mom walked in and screamed. "What are you thinking, George? You bought him one of those evil dolls!"

I reached out and grabbed the Herobrine from my dad. "It's not a doll, Mom. It's an action figure."

"Oh come on, Lois. Chill out," said my dad. "It's just a toy."

My mom shook her head back and forth rapidly. "It is not a toy. It's a depiction of pure evil. Besides, even if it were 'just a toy,' why would you want your son to have such a horrible thing?"

"Oh, all the kids want these things and none of them have them. Sometimes it's good to get kids things even if you don't think they should have them. Just to see how they can handle it."

"Yeah, Mom, I can handle it," I said in my own defense.

My mom reached for the Talking Herobrine to snatch it from my hands. I pulled away from her. "No. I get to keep it," I whined.

My mom relented with a sigh. "Okay, you can keep it for now. But if I see you playing with it wrong or if you start acting all evil or something, I'm going to take it away. Deal?"

I reached out and shook my mom's hand. "Deal."

"Now that that is settled," said my dad, "push that button on his back. That's how he talks. Must be some sort of redstone contraption inside of there."

"Cool," I said as I reached out and pressed the button.

"My name is Talking Herobrine, and I love you," said the doll in an eerie, high-pitched voice.

"Awesome!" I said, as my dad smiled at me. I looked over at my mom with a big grin on my face as she just stood there with a frown, shivering because she was so afraid of the toy.

"Do it again," said my dad.

I pushed the button again. "My name is Talking Herobrine, and I want to play with you."

"And I want to play a game with you too," I said. "Let's go play" – I looked at my mom – "err, do homework."

I rushed into my room and put Talking Herobrine on the desk next to my math homework.

"So, Talking Herobrine, what do you think about math?" I reached out and pushed the button on his back.

"My name is Talking Herobrine, and I like you."

I guess it doesn't engage in conversations.

Well, I had to finish my homework before I could play outside with my friends. I wanted to show them my new Talking Herobrine, so I quickly finished my math homework, grabbed my new toy, said goodbye to my parents and went outside.

I was able to find my friends, Cindy and Joe, playing in Joe's front yard. They were building some sort of dirt hut.

"Hey guys," I said. "Check it out!"

They looked up at me and, for a moment their eyes darted around wondering what I wanted them to check out. Then they both saw it at the same time.

"No way!" said Joe.

"Sick!" said Cindy.

"How did you get a Talking Herobrine?" asked Joe as he walked up to me followed closely by Cindy.

"My dad just brought it home today for no reason."

"Dude, you have the coolest dad ever. My parents won't let me have one of those things. They think they are evil," said Cindy with a frown.

I nodded. "Yeah, my mom didn't want me to have it. She said if I start acting all evil, she was going to take it away."

Joe laughed. "How can a toy make you be all evil?"

I shrugged. " I have no idea. Anyway, try it out," I said handing the toy to Cindy. “Press the button on its back.”

Cindy took Talking Herobrine, looked at it for a moment, and then pressed the button.

“My name is Talking Herobrine, and I don’t know you.”

“Whoa,” I said. “It’s like he actually was talking to you.”

Cindy laughed slightly uncomfortably. “Yeah right.” She pressed the button again.

“My name is Talking Herobrine, and I don’t want to play with you.”

Cindy got a strange look on her face and then handed Talking Herobrine to Joe.

"That's pretty weird," said Joe. "Let me give this a try." Joe pushed the button.

"My name is Talking Herobrine, and I don't know you either."

What the Notch?

"Dude, this thing is crazy," said Joe. "I'm going to push it again."

"My name is Talking Herobrine, and I don't like you one bit." Joe was in shock and dropped Talking Herobrine onto the ground.

I reached down and picked up my toy and brushed the dirt off it. "Come on, bro, don't get dirt on it."

Cindy shivered and said, "Dude, maybe your mom is right. Maybe that thing is evil."

I shook my head. "No way. Those are just random sayings that are programmed in there. Watch. I'll push the button." I pushed the button.

"My name is Talking Herobrine, and you had better never let me go."

Okay, I had to admit I was getting a little freaked out myself. I'm not gonna say that the toy was evil because, after all, it was just a toy. But, it was saying some pretty freaky stuff. If my mom overheard this, she would take it away in a heartbeat.

"I think you'd better leave," said Joe. Cindy nodded her head in agreement.

I sighed. "Come on,guys. We don't have to play with Talking Herobrine. Let's do something else."

At that moment I felt a slight vibration coming from the toy. I looked down and saw his eyes glowing!

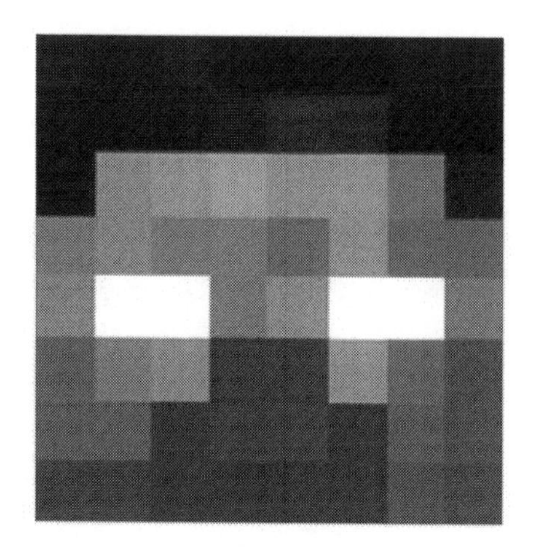

"Oh my Notch! Look!" I took the doll and showed it to Cindy and Joe. They looked at it, but they didn't see anything unusual.

"What, dude? It just looks like a stupid Herobrine toy," said Joe.

"But it's eyes were glowing. It vibrated, and then it ... it ... it looked at me."

Cindy looked me up and down with concern. "Either you are going crazy or that doll is evil. You need to get rid of that thing."

I couldn't get rid of Talking Herobrine. He was the most awesome toy ever, and no one else had one. Still, he was saying some strange things and it had felt like he was almost alive when he looked at me with his glowing eyes. I was really conflicted and didn't know what to do.

"Guys, I guess I'll just go home and eat some dinner. See you later."

Joe and Cindy waved at me as I left. I held Talking Herobrine by one arm limply at my side.

When I got home, my mom could tell something was wrong. "What is it, Harry?"

"Nothing, Mom," I said as I sat down at the table. "Just kind of got in an argument with a couple of my friends."

My mom sat down at the table and held my hand. She avoided looking at Talking Herobrine.

"What were you arguing about?"

"Uh, nothing. We just couldn't decide what game to play."

My mom smiled and patted my hand. "That happens sometimes. I'm sure tomorrow everything will be back to normal."

I glanced down at Talking Herobrine and thought I could see a faint glow in his eyes. "Yeah, I hope so."

I picked up Talking Herobrine and walked into my bedroom. I put him on my desk next to my math book. I stared at his eyes, almost hoping they would glow again so that I could confirm what I had seen and that I was not going crazy.

But, nothing happened.

A few minutes later, my mom called me for dinner. It was pretty good. Some mushrooms stew, a pork chop, some boiled carrots, and a slice of apple pie for dessert.

After dinner I went back to my room, but I didn't see Talking Herobrine. I knew I left them on my desk but now he wasn't there.

I went back out to the kitchen. "Dad, have you seen Talking Herobrine?"

He shook his head. “Don’t tell me you lost that thing already? That cost a lot of emeralds.”

“No, I left him on my desk before dinner and now he’s not there.” I turned to my mom who was doing the dishes. “What about you, Mom? Have you seen him?”

My mom hesitated for just a moment and then said, “The last I saw that thing was when you took it into your room.”

I scratched my head. “I just don’t understand this. How can he just disappear? It’s not like he can get up and walk away.”

I went back into my room and found a book to read. It was a

Minecraft diary by this author named Dr. Block. I thought his stuff was pretty good.

I read for about a couple of hours and was about halfway through the book I was reading, when I suddenly remembered something. During dinner, my mom had gotten up to use bathroom. And she had walked by my room to get there. I wondered if maybe she had stopped into my room and grabbed Talking Herobrine. I didn't want to think my mom would take it without telling me, but she really did think it was evil.

If she took it where would she put it?

I got up and went to the back door of our house. I went outside to the place we kept our garbage cans. We had two garbage cans. I opened the lid to the first one, and it was empty. I opened the lid of the second one and it was full of all kinds of disgusting old food and papers and things like that. It smelled horrid.

I was just about to put the lid back on the garbage can when I saw a dim glowing light coming from the bottom of the trashcan.

"Talking Herobrine? Is that you?"

I pushed the trashcan over spilling its nasty disgusting

contents on the ground. It smelled horrible, like a bowl of mushroom stew someone had left outside in the sun for like a week. I kicked to the pile of scummy trash with my feet and there he was, Talking Herobrine, stained with garbage. His eyes glowing with a fierce white light.

I picked him up and flicked all the nasty debris off of him as best I could and then pressed the button on his back.

"My name is Talking Herobrine, and I am very, very angry."

"I didn't do anything."

Herobrine's eyes glowed and stared at me. It was like he was ordering me to push the button. I pushed it.

"My name is Talking Herobrine, and you are going to do exactly what I say."

I dropped the toy back into the pile of garbage. This was beyond creepy, and I didn't want to deal with it. As I was walking away from the pile of garbage, I heard, "My name is Talking Herobrine, and you are going to turn around."

He was right. I turned around.

There, standing amidst the pile of garbage was Talking Herobrine. His eyes were glowing, and one of his arms was pointing directly at me.

"What? What do you want?" I said with through chattering teeth. I was surprised I could even form sounds, much less speech, because I was so scared.

"Take me to your mother. I need to tell her something."

I was powerless to refuse. I reached down and picked up Talking Herobrine. I could feel the life coursing through his tiny body. He was no toy. He was Herobrine, or at least he was a miniaturized version of Herobrine. Or something. I'm not sure exactly what he was really.

I slowly began to walk into the house carrying the stinky, garbage-covered Talking Herobrine. I walked into the kitchen and held him in front of me.

My mom was washing the dishes. When she turned around she screamed and dropped a dish. It shattered on the floor.

Talking Herobrine was staring at her with his glowing eyes.

"Get that thing out of here!" my mom yelled.

I could not move.

Talking Herobrine began to speak. "My name is Talking Herobrine, and you put me in the garbage. I don't like you. I am going to kill you."

I couldn't let this happen. I could feel Herobrine controlling my mind, but I wasn't going to let him kill my mom.

With all my strength, I threw Talking Herobrine onto the floor. And, then I began to stomp on him with my foot as hard as I could.

"You will regret this, boy," he said between stomps.

I stomped and stomped until he was nothing but a shattered mass of pebbles and splinters. And then, a very strange thing happened. The pile of debris disappeared in a puff of smoke and dropped a small obsidian star.

“Oh my Notch above,” gasped my mom. “It was alive!”

I turned away from the disgusting obsidian remains of the Talking Herobrine and I gave my mom a big hug. “You were right, Mom.”

At that moment, my dad walked in holding a book and eating a cookie. “What’s all the ruckus? Can’t you keep it down? I’m trying to read here.”

The Haunted Pork Chop

One night, I woke up a little after 2:00 a.m.. I tossed and turned in bed for a few minutes, trying to go back to sleep. But, there is no denying it, I was starving.

I got out of bed and walked down the hallway of my house. I was careful not to wake my mother and father and my little baby sister. I crept along the hallway, making sure that my footsteps were as silent as death, and made my way into the kitchen.

"What do I want to eat?" I whispered to myself as I rubbed my grumbling stomach.

I knew I could eat a cookie or some bread or some cold soup. My parents would not let me use the furnace to heat anything up. They thought I was still too young to be trusted around fire, but I was an eight-year-old villager. I totally knew what I was doing.

I decided to eat a couple of cookies and drink a glass of milk. I opened the cookie chest and pulled out two cookies and put them on a plate. Then I found a cup and poured some milk into it. I sat at the table and slowly chewed a cookie, trying to be as quiet as possible.

I finished my first cookie and was just about to bite into the second cookie when I saw something that made my blood run as cold as ice.

I mean I literally had the cookie in between my upper and lower teeth and was just about to bite into it, but the thing that I saw was so freaky that I was paralyzed with fear.

"Can I have a bite of that cookie?" it asked.

I slowly removed the cookie from my mouth and set it down on the plate. "Sure, do whatever you want," I said with a voice barely louder than a whisper.

It came towards me making a disgusting wet *GAH-lop* with every step it took. Yet it was floating in the air at the same time. It had no legs, but moved and sounded like it was walking.

Could it really be what I thought it was?

It reached down and picked up the cookie. Or, should I say some sort of invisible force reached down and lifted the cookie up towards its mouth. It was as if it had invisible hands and arms. It opened its mouth and took a satisfying bite of the cookie.

"Yummy. I love cookies," it said as it smacked its gooey lips together.

“I don’t want to be rude but are you a...?” I couldn’t finish my question. It seemed too insane to even ask, but I had to know.

It nodded what appeared to be its head up and down, making a sloshing squishing noise as it did so. “Yes, I am a pork chop.”

I felt the blood drain from my cube-shaped head. “But how are you talking? And how are you eating? Don’t pork chops normally just kind of sit there and wait to get eaten?”

The pork chop nodded its head. “Oh yes. I’ve already been eaten. You see, I am a ghost pork chop.”

A bolt of fear shot up my spine and exploded out the top of my head. Well, it didn't really *explode* out of my head or else I would've died. What I mean is I basically felt super scared and my head felt really stressed out and I got a bunch of goose bumps and stuff.

"So, why ... why ... why are you in my house right now?"

The ghostly pork chop looked at me and smiled, bits of cookie crumbs and chocolate chips stuck to its fang-like teeth. "Because, you ate me last night."

Night of the Living Zombie

I was sitting around in my room with my friends Bob and Rick. We were trying to think of something to do.

"I know, let's go fishing," said Bob.

"Fishing is boring," I said.

"Oh come on, Jeff, fishing is pretty fun," said Rick.

"We always do the same stuff, fishing, looking for apples, throwing rocks at squids, building dirt forts, punching trees. Ugh, I'm getting sick of it," I said.

"Well," said Bob with a suspicious grin on his face, "there is something we could try that we've never done before."

I was intrigued. I could tell Rick was interested to. After all, we were twelve-year-old villagers, and we were ready for anything.

"What?" I asked.

"Well, I found an interesting book tucked away in my parents' library. It looks really old and has all sorts of weird potion recipes in it. There was one that was really extreme," said Bob.

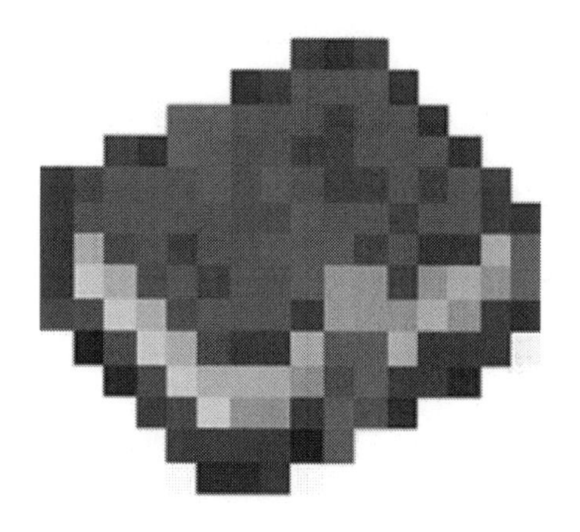

"What? Tell us," Rick and I said in unison.

Bob's eyes darted to the left and the right checking to make sure we were alone. Of course, this was pointless and stupid because we were in my room and the door was shut and no one else was in there with us.

"It was a potion that turns you into a zombie for one night," he whispered.

I was shocked. *Was it actually possible that a potion existed that could turn you into a zombie for a night?!?* This sounded pretty radical and I wasn't sure it was the best idea, but I needed more information.

"Really? What's in the potion?" I asked.

"It's actually a pretty simple potion," said Bob. "You only need some rotten flesh, ink from a squid, and a chorus fruit."

"Simple?!?" I said. "I mean, squid ink is pretty easy to get and

I'm sure we could make rotten flesh just by leaving some meat out in the sun for a few days, but how are we supposed to get chorus fruit? That stuff is super rare."

"Actually, my dad has a few chorus fruits. He traded a player for them a few weeks ago. He's been saving them for a special occasion," said Rick.

Bob's eyes opened very wide. "Dude, we only need one chorus fruit for this potion. Do you think you could sneak one?"

Rick shrugged. "I'm sure I could. I don't really like the idea of stealing from my dad though."

"Oh come on, Rick," I said, begging. "This will be so awesome if we could be zombies for a day. All of our friends will be totally jelly."

Bob slapped his head. "Did you just say 'jelly' to mean jealous?"

"So?"

"What are you, a three-year-old?"

Rick ignored our banter, lost in thought. Finally, he relented. "Okay. I'll see what I can do." Then Rick smiled briefly. "Yeah, it would be pretty awesome to be a zombie for a day."

***

A few days later, after Rick had acquired one of the chorus fruit and after we had a left piece of cow flesh in the sun for a few days to rot, the three of us met at the fort we had been building over the past few days.

We found an isolated spot in the nearby forest. We had dug a hole several blocks deep and then made a roof out of wooden planks which we then covered with dirt blocks. It looked just like a field from the outside. We had carved an entry tunnel underground to the bottom of the hill and blocked it with some large boulders. It was the ultimate secret fort. Inside of our fort, we had a few chairs and a brewing stand.

Now that we had all the ingredients, Bob pulled out his book with the recipe for the zombie potion. We followed it precisely. It took a while for the brewing to occur but once it was done there was enough potion for the three of us.

We looked at the dark black, syrupy and horrifically stinky potion. We were very somber. I was getting nervous. I'm sure my friends were too.

"Well, should we do it?" asked Bob, sounding unsure for the first time during this entire process.

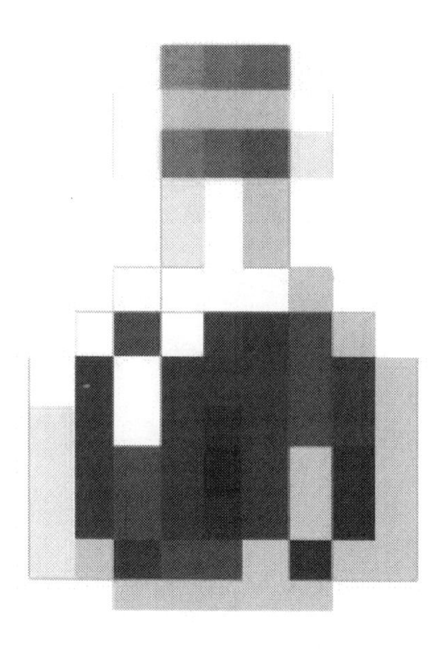

"I don't know, this seemed like a good idea before but now that we really have the potion, I'm not so sure," muttered Rick.

"I think we should still do it, but it's still daytime right now. Won't we burn if we turn ourselves into zombies and go out into the daylight?" I asked.

“Let me consult book,” said Bob.

Bob read the handful of paragraphs following the recipe instructions in the book. There were some warnings about not using the potion more than twice a year and being careful not to drink more than the recommended dosage, but nothing about whether you would burn or not.

“Well, every zombie I’ve ever seen exposed to sunlight burns to death. I’m not gonna risk turning into a zombie during the daytime. And, I want to make sure that I have as much time as possible to wander around in my zombie form, so I will wait until after dark to drink it," I said.

My friends agreed. We left our fort and walked around outside in the sun and enjoyed not being zombies.

"Man, it would be so lame to not be able the walk around in the sunshine," said Rick.

"Totally. Those zombies really have a raw deal," I said.

Bob smiled. "You mean a rotten deal."

***

We went back to our houses and ate early dinners. I told my parents I was going to spend the night at Bob's house. Bob and Rick told similar stories to their parents. Our whereabouts sufficiently concealed, we returned to our fort to arrive there before darkness fell and the real zombies emerged.

We stood in the fort staring at each other. Each of us held a dose of potion in a bottle in our hands. The atmosphere was tense, but electric with excitement.

"It must be dark by now," said Bob. "Let's do it."

We each slowly raised the vials of zombie potion to our mouths. They smelled awful.

"I don't know if I can drink this. It's so stinky," I said, holding back an urge to vomit.

“Just pinch your nose shut with your fingers. Then you won’t be able to smell,” said Bob.

I pinched my nose shut and saw Bob and Rick do the same. We each quickly raised our bottles to our lips and gulped down the potion.

The taste was disgusting, like drinking sulfur mixed with decay mixed with something for which I have no words. I imagined it must be what it would taste like to drink death, if death were a liquid you could put in a bottle.

At first, nothing happened. Within a few seconds, however, I had a horrible cramp in my stomach and fell to the ground clutching my gut. “I think I’m going to die,” I yelled.

Bob and Rick also fell to the ground clutching their guts and moaning in agony.

“What did you do to us, Bob?” asked Rick.

“I followed the recipe. I swear!” yelled Bob.

After that, I passed out.

***

When I regained consciousness, I felt much better. The cramping in my stomach was gone, as was the disgusting taste in my mouth. I was lying on my back in the fort, looking up at the wood plank roof. I did not feel any different.

I sighed. “I guess the potion didn’t work.”

I sat up and ... saw two young zombies sitting in front of me!

I was chilled to my core.

“Is that you Bob? Rick?”

They pumped their fists. “Yeah, isn’t awesome!” they said in unison.

I looked down at my arms and saw that they were a sickly green color. I was still wearing my villager robe, but otherwise I looked like a zombie.

“This is so sick!” I said, flexing my arms like a bodybuilder.

“Yeah, we should go into the village and scare people,” said Rick, smiling wide and exposing his rotten zombie teeth.

I laughed. “Yeah, that’s a good idea. Let’s start with my little sister,” I said, rubbing my hands together gleefully.

We walked to the exit of our fort and emerged into the night. We could see a few skeletons and zombies wandering around, but we weren't scared. They would think we were just like them.

"Let's test out how good this potion really is," I said. "Follow me."

We walked slowly over to a couple of zombies wandering around and moaning. When we got relatively close to them, we all started to walk in shambling zombie style, leaning slightly to one side with our arms outstretched and moaning and groaning. We walked right up to the other zombies and they didn't even give us a second look.

When we were far enough away so that the zombies couldn't hear us, I said, "That was so awesome."

We all gave each other high fives to congratulate each other on our complete dominance.

"Okay guys, let's go scare everyone in the village."

It was pretty fun scaring my sister, our parents and our friends. You can imagine the hilarity, I am sure.

But, we did make one miscalculation, and that is that when zombies come into the village and there's a player around, the player tries to kill the zombies. And that's what happened with us.

We were just about to go into the house of one of Rick's next-door neighbors to scare him when a player showed up out of nowhere.

The player didn't have any armor and his only weapon was a stick, so he was clearly a pathetic noob. Still, getting hit with a stick and punched a few times hurts a lot. We probably could've killed the player if we really wanted to but instead we ran away, leaving the player scratching his head because he probably never had seen zombies run as fast as we could.

After our encounter with the wimpy player, we decided that it would probably be best just to walk around the forest and stay close to our fort so that we could duck inside when the daylight came, if we hadn't changed back to ourselves by then. According to Bob, the potion was supposed to last between eight and twelve hours.

We spent the rest of the night wandering around looking at strange sights we never would have seen as simple villagers. In fact, we walked up to a skeleton and just reached out and touched his rib cage. He looked at us like we were crazy and held up his bow and arrow menacingly, threatening to shoot us. We backed away after that.

From then on, we just pretended to accidentally bump into skeletons or spiders just so we could feel what they felt like. It was going to be so awesome to be able to tell our friends that we had touched skeletons and spiders. There would be so jealous.

When we could see the first light of the false dawn glowing dimly in the east, we walked back to our fortress and went inside to shield our zombie bodies from the rays of the sun.

None of us had yet changed back into our villager forms, but we were all very tired.

"Guys, I think I'm going to go to sleep. I'll probably wake up after the zombie potion wears off," I said, yawning.

"Okay," said Bob. "I'm too excited to go to sleep. I am going to stay up and watch my flesh turn back to regular villager flesh. It'll be awesome."

"Yeah, I think I'll do that too," said Rick.

I yawned again, the stink of my rotten breath surprisingly pungent. "Okay, wake me up when it happens."

***

I was being shaken roughly. At first, I thought I was just having a dream, but then I realized it was Rick and Bob shaking me.

"Wake up, Jeff! Wake up!" They were saying. There was a hint of desperation in their voices.

I shook my head and rubbed my eyes. I was still very groggy but was able to manage, "Stop shoving me, you jerks!"

"Something went wrong!" said Bob, a look of terror on his face.

I looked at Bob and Rick who were back to their normal villager selves. "What do you mean, something went wrong? You guys look fine."

There was silence. And that's when I began to realize that they weren't worried about themselves. They were worried about me.

I looked down at my arms and they were still a sickly green shade of zombie. I screamed, "No!"

"I don't know what happened. Rick and I started to change back and we went into convulsions of pain and passed out. When we woke up, you were still a zombie," said Bob.

"What can I do?" I asked desperately. "I can't stay a zombie for the rest of my life."

I looked at Bob and saw something in his eyes. It was like he knew something. Something that he needed to tell me but was too scared.

“Tell me.”

Bob sighed heavily. “When Rick and I woke up and realized that you are still a zombie, I read through the book to see if there was an explanation as to why. There was.” Bob stopped talking and shook his head. A tear rolled down his cheek.

My heart was racing. *What could it be?*

“Tell me, Bob.”

"It wasn't on the same page as the recipe for the zombie potion. It was in the back. In the appendix. There is a section of general warnings."

Bob paused. The tears now were streaming down his face. He heaved a sigh and finished. "One of the warnings says that you must be awake to transform back to your normal form or … or … you never will."

The Telltale Emerald

My name is Bartoli, but people call me "Bart."

I live in the village of Zombie Bane. It is kind of in the middle of nowhere, but is a popular tourist destination because many years ago another villager killed a bunch of zombies all by himself.

Anyway, villagers and players come from all over Minecraft in order to see where the miraculous zombie massacre occurred. And, best of all, they like to buy things.

I run a small shop where tourists can buy snacks and souvenirs. I earn quite a few emeralds every day, but I am not rich. Not compared to most of the villagers in Zombie Bane.

That upsets me.

It is not fair.

Every day when I come home from the shop, my wife is upset because I didn't make enough money.

"Hurrr," she says. "Mrs. Johnson just told me that her husband made 500 emeralds today. How many did you make?"

"Hurrr, only 127," I sigh.

She turns up her thick, rectangular nose. "And you call yourself a villager?"

My children constantly want me to give them emeralds to spend on things like blank diaries to write in, cookies, and cupcakes.

I am sick and tired of my wife's disdain and my children constantly spending all of my money. So, today, I did something amazing: I stole an emerald from a customer!

Just one.

I just wanted to see if I could do it.

And, I did.

The customer and his family, his wife and eight children, came into my shop. They were distracted. They were in a hurry.

"Hurrr," the customer said to his family, "grab something to eat. We need to get to the zombie massacre reenactment. It starts in two minutes."

The family grabbed snacks and dropped them on the counter in front of me.

"That will be, hurrr, 13 emeralds," I told the man after I totalled everything. And, that was the true, authentic, accurate, no-kidding price: 13 emeralds.

The man reached into his robe and pulled out a handful of emeralds and dropped them on the counter. "Take what you need," he said, grabbing the pile of snacks and turning his back to me.

While he passed out the snacks to his family and was distracted, I took 14 emeralds and put them in my emerald chest.

If he notices, I'll just tell him I made a mistake.

I began to sweat with excitement. I knew it was wrong to steal, but I didn't care. I'd show everyone that I could be a rich villager too!

"Hurrr, your change, sir," I said, pointing to the 3 emeralds remaining on the counter.

The man quickly grabbed the emeralds and stuffed them back into his robe. "Thank you," he said to me. Then turned to his family and demanded, "Hurry up, we are going to miss the zombie slaughter!"

The family cheered and then left my shop.

I had done it. I had stolen an emerald.

I felt empowered.

I felt awesome.

I now knew how I would become rich.

I would become a thief.

But, no one could know. I could not tell my wife or family. I could not tell my friends.

Yes, it will remain a secret.

I was so excited that I could no longer concentrate on my work. I closed my shop and went home early. I brought my stolen emerald with me so that I could admire it.

No one was home when I arrived.

Good. I will be able to spend some quality alone time with my emerald.

I went into the kitchen and put the emerald on the table. I sat down next to it. Oh, how it sparkled! I loved my new emerald. I loved stealing it.

I looked at it for quite a while. It was as if I were seeing an emerald for the first time in my life. And, I was so happy.

But, then, I thought about my family. They would be home soon. They would ask me where I got the emerald. I could not tell them I stole it. I had to hide it somewhere. But, where?

In a chest?

In a box?

In a nest?

Under rocks?

No, none of those places would do. I needed somewhere more secure. Somewhere that they would never look. Somewhere big enough to hide all of the emeralds I was going to steal.

But, I could think of no place that would do. I became so frustrated that I literally began to bang my head against the wall.

Thwack!

Thwack!

And, that was when I realized that I could stash the emerald inside the wall of my house. No one would ever suspect that the wooden walls of my humble dwelling were actually swelling with emeralds.

"Bwahahaha!" I laughed like a maniac.

I punched the wall of the kitchen and opened up a small hole.

I stuffed the emerald inside and then patched the wall.

Like it never happened.

My family came home about thirty minutes later.

“What are you doing home so early?” asked my wife. “Shouldn’t you be earning emeralds?”

I smiled. “Sorry, I felt a little sick earlier and decided to close the shop for the day. I’ll be back at it first thing tomorrow.”

She squinted her eyes at me. “You don’t look sick.”

I rolled my eyes. “Well, I am.”

“Are you gonna barf?” asked my young son.

"I don't think so," I said, placing my hand on my stomach and pretending to be nauseous.

It was then that we heard a knock on the door. My wife answered it and returned. "Some one's here to see you. It's a police officer."

I suddenly did feel very sick. Nauseous even.

"Did he say what he wanted?" I asked.

My wife shook her head. "It's a 'she' actually. And, no, she did not say what she wanted."

I stood up and walked toward the front of the house. As I did, I thought I noticed a green glow coming from the spot where I had stashed the emerald.

What the Notch...?

I shook my head and the glow disappeared.

I left the kitchen and walked to the front door. “Yes, officer? How can I help you?”

The officer was dressed in a dark blue uniform and wore reflective sunglasses so that instead of her eyes, all I could see were miniature reflections of my own face.

“You own that knickknack shop on Second and Oak?”

I nodded. “Yes.”

"Following up on a report of a stolen emerald."

I suddenly felt faint. "You don't think I ...?"

She laughed. "Oh, no, sir. It is just that the person who claims to have had the emerald stolen visited a few different shops, and one was yours. Just wondering if you saw anything unusual."

I felt much better. "No, nothing unusual."

But, I was seeing something unusual, right now. I could see a green glow reflecting in the mirrored sunglasses worn by the officer.

I snapped my head around and looked behind me. No glow. I snapped my head forward again and the glow was there.

Was I going mad?

The officer noticed my strange behavior. “You alright, sir?”

I was beginning to sweat. “Uh, yeah, just feeling a little under the weather,” I said, tugging at the collar of my robe to let in some cool air.

“I see,” said the officer with a hint of suspicion in her voice. The officer took out a book and a quill. “Just a few more questions. I noticed you closed your shop early today. Why is that?”

She knows something. She suspects. Maybe she sees the glow and is toying with me.

"Uh, felt sick."

"Mmm. You just mentioned that and, well, you don't look so good, sir," she said. "Did you notice anyone suspicious lurking in or near your shop today?"

The green glow had spread from the officer's glasses to her nose and cheeks.

What the Notch is happening?

"Lurking? Uh, no. I haven't. I mean I didn't see. No. No lurking," I stammered.

The officer wrote down some notes in the book. "Sir, you really don't look good."

The glow had spread further. The officer's entire face was now green.

This. Is. Not. Happening.

"Come to me."

"What?" I said.

"I didn't say anything," said the officer.

"Come to me. Look at me."

Was the emerald calling to me?

"You took me. Now, worship me!"

"I won't," I screamed.

"Sir, are you alright. You look a little ... um ... green."

I shrieked. “It’s the emerald. It has made me green.”

I bolted into the house. The officer followed behind me. I looked back at her. She had pulled her police-issued, military-grade diamond sword from her inventory.

She’s after me. I have to get there first.

I ran past my wife and children. They looked at me like I was some sort of freakish ghost. Like they did not even recognized me.

I had to get to the emerald before I turned completely green. Before it took over my body and corrupted me forever.

I rushed to the spot in the wall and punched it as hard as I could. Over and over, with all my force, I punched the wall until it disintegrated.

"Sir, calm down," the officer said, taking a defensive stance and holding her sword in front of her.

I looked into the wall and saw the emerald. It pulsed with evil green light. I grabbed it and tossed it on the floor in front of the police officer.

"There. I took it. I stole the emerald. I don't care. Take it away. It is trying to control my mind."

The ambulance arrived a few minutes later. I was placed in a straight jacket and driven to a big white building surrounded by an expansive green lawn.

They put me in a white padded room. It had a window.

But, when I looked out all I could see was the lawn. A massive lawn. A lawn that stretched to the horizon.

A dark *green* lawn.

The Glare Witch Project

The day before Halloween, I was hanging out with my two best friends.

Wait a minute, let me back up.

My name is Justin, and I'm a 12-year-old Minecraft villager. This is my diary. And I'm about to write down one of the scariest things that ever happened to me ever.

If you're reading this diary, first of all, you should stop because it's private. But if you insist on continuing, just know that what I'm about to write was so terrifying the you might have nightmares for months.

Seriously, if you are a scaredy-cat, DO NOT READ this!

Anyway, as I was saying, I was hanging out with my two best friends, Jim and Mike. It was the day before Halloween, and we were talking about spooky stuff.

"Did you know that on Halloween zombies sometimes have pumpkin heads?" said Mike, his mouth hanging open with awe.

I rolled my eyes. “Dude. Everybody knows that. What are you, like, three years old?”

Mike punched me in the arm. “Don’t make fun of me.”

“Ouch!” I said as I rubbed my shoulder. It really hurt. Mike must’ve been practicing punching trees or something because he had never hit me this hard before.

“You guys are both such babies,” said Jim. “Who cares about zombies. They’re so wimpy.”

“I don’t think zombies are very wimpy,” I said. “I mean, how many zombies have you killed?”

Jim shrugged. “I’ve killed a few zombies.”

I laughed. "No way."

"Yeah, since when have you been a zombie killer? I bet you don't even know how to use a sword," said Mike.

Jim started to turn red. He arched his eyebrows and wrinkled his forehead. "Well, I have too killed zombies, and I don't care what you say. Besides, like I said, zombies are wimpy and lame."

"Well what's more dominant than zombies?" asked Mike.

"Herobrine, obviously," I said, proud of myself.

"Herobrine is old news. I know about something that's way more scary than Herobrine," said Jim.

I didn't believe him. *What could be more dominant and scary than Herobrine?*

Mike gasped. "Are you talking about Entity 303?"

Jim laughed again. "Entity 303? You believe that nonsense? No, I'm talking about the Glare Witch."

I'd never heard of the Glare Witch before. It sounded kind of strange. "What are you talking about? What's the Glare Witch?"

Jim sat down on a tree stump. "Sit down and I'll tell you. You shouldn't be standing up when you hear this."

What a drama queen!

Mike and I both sat down. Jim began to speak in a low voice, to be sure no one would overhear what he said. "It is said that the Glare Witch lives somewhere deep within the forest behind the village. Supposedly, she has a hut out there where she kidnaps villagers and does who-knows-what to them."

I'll admit, that sounded pretty creepy. However, I'd been in that forest a bunch of times with my dad and my friends, and never saw anything that would've made me think there was a witch living in there.

Jim continued. "The thing is, the Glare Witch is not like any other witch. Most witches just sit in their huts making potions. This witch doesn't need potions. They say, she sneaks up behind you and then when you turn around she glares at you in the face. When that happens you instantly die and respawn as a silverfish!"

As scary as the story sounded, I didn't want to let Jim know he had frightened me. "That's the most ridiculous thing I've ever heard. How can a witch kill you and make you respawn as a silverfish just by looking at you?"

"Yeah, that sounds stupid," said Mike through chattering teeth.

My bro was way more scared than I.

“Well, that’s what I heard. And, I don’t believe it either,” said Jim. “I say, we go explore the forest and look for the Glare Witch. When we find nothing, we can come back to the village and tell everyone that the story is not true.”

“What you mean tell everyone? Who else knows about the Glare Witch?” I asked.

“All the adults know,” said Jim. “That’s how I found out about it. I overheard my dad talking to another adult. Apparently, when you turn 18 and become an adult, they all tell you this secret.”

I guess that sounded possible. Adults were always sneaking around and keeping secrets from kids, so it would make since they would keep a secret as terrifying as this. But, if it were true, then why didn't my parents keep me away from the forest?

"Okay, Jim, I'll go to the forest with you," I said, fairly confident the Glare Witch story was a bunch of nonsense.

"Are you coming too, Mike," asked Jim.

I could tell Mike didn't want to come, but he was a joiner. If the Jim and I did something, Mike had

had to come with us just because he did want to be left out.

"Yeah, I guess I'll come with you."

I went back to my house and gathered supplies. I packed some cookies and milk in my inventory. I packed about two dozen cookies because I thought we might be out all night. If we didn't find the Glare Witch by morning, I was going to give up.

I packed a couple of torches from my parents' torch supply and put my child-sized iron sword in my inventory. There would likely be zombies lurking around in the forest, so just in case I wanted to be prepared.

That night, I waited until my parents and little brother went to sleep before I snuck out of the house. I felt bad about sneaking out, but this was an adventure. And I didn't want Jim to think that I was too scared to go.

A few minutes later I met Jim and Mike at the rendezvous spot.

"Did you bring some food?" asked Jim.

"Yes."

"Did you bring some torches?"

"Yes."

"Did you bring some weapons?"

"Yes. I brought a sword," I said, pulling the sword out and swishing it through the air.

Jim laughed. "That's a pretty small sword, dude."

I was little embarrassed. "It's the only thing I have."

"It's cool. I brought three full-size swords, one for each of us. Here," said Jim as he pulled two of the swords out of his inventory and handed one each to Mike and me. Mike looked very nervous holding the sword.

Jim grinned. "I also have a bow and arrows. We might need to shoot some flaming arrows at zombies if too many approach us at any one time."

"I'm not sure I want to do this anymore," whined Mike.

"Don't be a wimpy villager," said Jim.

"Leave him alone, Jim. If he doesn't to come, he doesn't have to," I said.

"Thanks, Justin, but I'll come," said Mike.

We ignited our torches and walked into the woods. At first, we heard nothing and saw nothing, then we began to hear rustling.

"What is that?" asked Mike, a quiver of fear in his voice.

"I don't know," I whispered.

We stood still and listened to the rustling. It got closer and closer. We were just about to pull out our iron swords when a tiny bunny rabbit hopped a few feet from us.

We all breathed a sigh of relief.

"Stupid rabbit," I said. I kicked the rabbit and it scurried off.

We walked for about ten minutes without anything happening. There was a half moon that provided us with some light but we still needed our torches to see where we were going. A few times we heard zombies moaning in the distance but none of them came towards us.

"Let's climb over that hill," suggested Mike, pointing. "If there is a witch's hut, maybe we can see some light coming from it."

"That's a good idea," I said.

We hiked up to the top of the nearby hill as Mike suggested.

When we got to the top, we thought we could see a light in the forest.

"Look," said Jim, pointing. "That might be a witch's hut or maybe a campfire of some player."

"I don't think a player would be so stupid as to make a fire in the middle of the night, although I've seen some pretty noobie noobs from time to time. I bet it *is* the witch's hut," I said.

Mike began to shake. "This is pretty freaky. What happens if we find the Glare Witch?"

"Don't look at her face. Duh," said Jim.

We began walking down the hill in the direction of the light.

When we were about halfway down the hill, Mike tripped over something -- was it a rabbit? -- and began to tumble downhill so quickly that we soon lost sight of him.

"Mike! Mike!" we yelled.

The only response was dead silence.

"Come on, maybe he hit his head and is knocked out," said Jim.

We ran down the hill in the direction Mike had rolled. He had tumbled very far so it took us about two minutes to get to the bottom of the hill, but we didn't find any sign of Mike.

Well, that's not entirely true. We found a drop pile containing one sword, some cookies, and Mike's diary.

"Oh my Notch, do you think the Glare Witch got him?"

Jim hung his head and sighed. "Something got him. This is a drop pile. He's dead."

"We need to go back to the village now," I said. I could feel hysterical fear beginning to creep into my brain. "If it's not the Glare Witch, then there's some sort of madman roaming the forest killing villagers. Either way, we need to leave."

"Should we really go?" asked Jim. "I mean, Mike is dead, and should we really run away and not try to find what or who did this to him?" said Jim.

"You're crazy. We need to go," I demanded.

"No. I'm staying and will keep looking for the Glare Witch. You can go if you want, you wimp."

I wanted to punch Jim in the face. But that would not have helped. I either needed to leave or stay. I did not want to be by myself in the dangerous forest, so, for better or worse, I decided to stay with Jim. If we were lucky, we would wander around all night, and find nothing. Then we could return to the village.

"Okay, I'll come with you, I guess."

We walked for a few more minutes, and then we came to a stream. We held our torches near the water and saw some fish idly swimming against the current. They looked like they were asleep.

"I'm hungry," I said as I pulled a couple of cookies and some milk from my inventory. The cookies tasted good. The sugar helped calm my nerves after the horrific discovery of Mike's drop pile.

After I finished eating the cookies, I looked over at Jim. "Any idea which direction we should go?"

Jim shrugged. "Let's follow the stream for a while. I think a witch would put her hut near a stream so she could get easy access to water for making potions and such."

That was actually a pretty good idea. "Which way do you want to go?"

"Let's go upstream. It will take us farther away from the village, and I would suspect that the witch would live as far from the villager she could."

Again, this was a good idea, though I did not want to be moving away from the village. Still, I had to admit that if we were hunting a witch, this was the best way to do it.

Or, maybe I was completely wrong?

"Okay, let's go that way," I said, pulling my sword from my inventory. "But if I see anything move, I am going to stab first and ask questions later."

"I agree," said Jim pulling his sword out as well.

We walked along the stream for about 10 minutes when all of a sudden a tree fell down right in between the two of us!

John backed up just in time to avoid being squished by the tree. I looked over the trunk of the tree at

Jim and said, "Wow, that was close."

Jim was panting. "Yeah, that *was* close. Let me climb over the tree and then we can keep going."

As Jim started to climb the tree, he slipped on something -- slimy fish scales? -- and fell into the water. The current of the stream, which had appeared so weak earlier, suddenly became very swift and pushed Jim along the river away from me.

"Jim! Jim! Swim to shore," I yelled. Jim was trying to swim to the shore, but wasn't having a good time of it. The river was moving so quickly that he disappeared around the bend in the river. I ran as fast as I could along the stream bank to catch up with him.

A couple minutes later I found him. Or, what was left of them.

It was his drop pile.

Now, I was really scared. Both of my friends were dead. I was alone in the middle the forest. And, I had no idea what had killed them.

"I need to get home," I whimpered to myself. "I need to get home."

I pulled two torches from my inventory and ignited both of them. I held one in each arm so that I could see very clearly where was going. I walked along the stream.

The slightest noise sent lightning bolts of fear up and down my spine.

I knew if I followed the stream I'll get to the village in about half an hour. I was walking as quickly as I could, but did not want to run for fear of making too much noise and drawing the attention of wandering zombies and skeletons.

I'd been walking for about 20 minutes and was within sight of the village. I could see tiny beams of light poking out of villager houses. I could even spot my own house, where my parents and little brother were sleeping without any knowledge that I did gone on a stupid quest or that two of my friends had died.

I paused for just a moment to catch my breath.

I set down one torch in order to pull out a cookie and eat it quickly.

I pulled out the milk jug and drank a couple of swallows to quench my thirst. I put it back in my inventory

I bent down to pick the torch off the ground and when I stood up I saw...

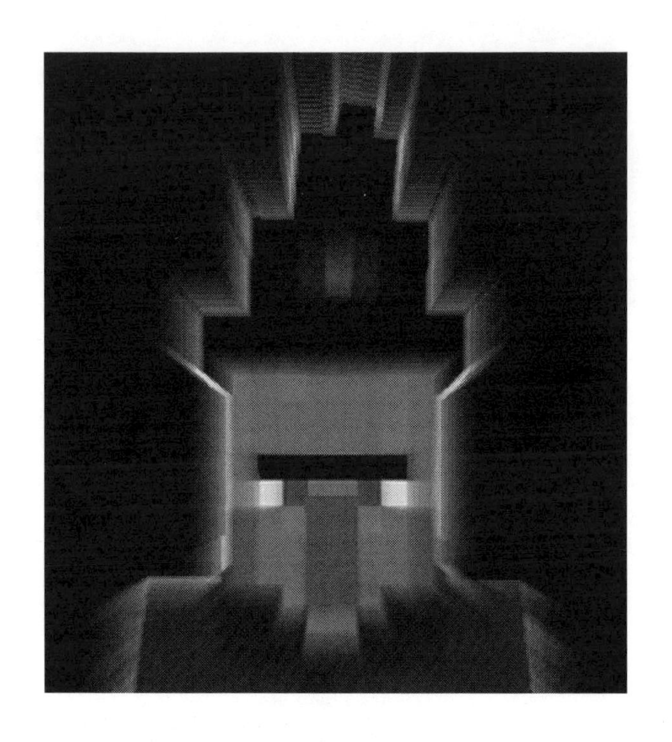

The End

Author's Note

I hope you enjoyed these spooky stories and that you weren't too scared.

If you liked these stories, **please leave a review** and let me know if you want more spooky Minecraft stories. I'm happy to write them if I know people want to read them.

Please be sure to stop by my website (www.drblockbooks.com), facebook (facebook.com/drblockbooks) or instagram (instagram.com/drblockbooks) to learn when the latest books become available!

See you soon,

Dr. Block

Made in the USA
Middletown, DE
12 October 2018